Higgs Discovery

Lisa Randall is Professor of Physics at Harvard University. She is one of today's most influential and most cited theoretical physicists, and has received numerous awards and honours. Her research in particle physics and cosmology has led to numerous testable speculations about physics beyond the Standard Model.

Randall is a member of the National Academy of Sciences, the American Philosophical Society, the American Academy of Arts and Sciences, the American Physical Society, an honorary member of the Institute of Physics, and is the recipient of several honorary degrees. She is the author of *Knocking on Heaven's Door* and *Warped Passages*.

When not solving the problems of the universe, she can be found rock-climbing, skiing, or contributing to art–science connections.

ALSO BY LISA RANDALL

Warped Passages: Unravelling the Universe's Hidden Dimensions

Knocking on Heaven's Door: How Physics and Scientific Thinking Illuminate the Universe and the Modern World

Higgs Discovery

THE POWER OF EMPTY SPACE

Lisa Randall

THE BODLEY HEAD
LONDON

Published by The Bodley Head 2012

2 4 6 8 10 9 7 5 3

First published in Great Britain in 2012 by
The Bodley Head
Random House, 20 Vauxhall Bridge Road,
London SW1V 2SA

www.bodleyhead.co.uk
www.vintage-books.co.uk

Addresses for companies within The Random House Group Limited can be found at:
www.randomhouse.co.uk/offices.htm

The Random House Group Limited Reg. No. 954009

A CIP catalogue record for this book
is available from the British Library

ISBN 9781847922571

The Random House Group Limited supports The Forest Stewardship Council (FSC®), the
leading international forest certification organisation. Our books carrying the FSC label are
printed on FSC® certified paper. FSC is the only forest certification scheme endorsed by
the leading environmental organisations, including Greenpeace. Our paper procurement
policy can be found at www.randomhouse.co.uk/environment

Printed and bound in Great Britain by Clays Ltd, St Ives plc

Contents

Preface: Higgs Discovery

On July 4, 2012, along with many other people around the globe who were glued to their computers, I learned that a new particle had been discovered at the Large Hadron Collider (LHC) near Geneva. In what is now a well-publicized but nonetheless stunning turn of events, spokespeople from CMS and ATLAS, the two major LHC experiments, announced that a particle related to the Higgs mechanism, whereby elementary particles acquire their masses, had been found. I was flabbergasted. This was actually a discovery, not a mere hint or partial evidence. Enough data had been collected to meet the rigorous standards that particle physics experiments maintain for claiming a new particle's existence. The accumulation and analysis of sufficient evidence was all the more impressive because the date of the announcement had been fixed in advance to coincide with a major international physics conference occurring in Australia that same week. And what was more exciting still was that the particle looks a lot like a particle called the Higgs boson.

A Higgs boson is not just a new particle, but a new type of particle. The thrill in this particular discovery was that it was not simply a confirmation of definite expectations. Unlike many particle discoveries in my physics lifetime, for which we pretty much knew in advance what had to exist, no physicist could guarantee that a Higgs boson would be found in the energy range that the experiments currently cover—or even found at all. Most thought something like a Higgs boson should be present in nature, but we didn't know with certainty that its properties would permit experiments to find it this year. In fact, some physicists, Stephen Hawking among them, lost bets when it was found.

This discovery confirms that the Standard Model of particle physics is consistent. The Standard Model describes the most elementary components that are known in matter, such as quarks, leptons (like the electron), and the three nongravitational forces through which they interact— electromagnetism, the weak nuclear force, and the strong nuclear force. Most Standard Model particles have nonzero masses, which we know through many measurements. The Standard Model including those masses gives completely consistent predictions for all known particle phenomena at the level of precision of a fraction of a percent.

But the origin of those particle masses was not yet known. If particles had mass from the get-go, the theory would have been inconsistent and would have made nonsensical predictions such as probabilities of energetic particles interacting that were greater than one. Some new ingredient was required to allow for those masses. That new ingredient is the Higgs mechanism, and the particle that was found is very likely the Higgs boson that signals the mechanism's existence and tells how it is implemented. With improved statistics, which is to say with more information after the experiments run longer, we will learn more about what underlies the Higgs mechanism and hence the Standard Model.

Though a discovery was indeed announced, it was in fact made with some of the caution I had come to expect from particle physics announcements. Because the measurements had identified barely enough Higgs boson events to claim a discovery, they certainly didn't yet have enough data to measure all the newly discovered particle's properties and interactions accurately enough to assure that it is a single Higgs boson with precisely the properties such a particle is expected to have. A deviation from expectations could turn out to be even more interesting than something in perfect accord with predictions. It would be conclusive evidence for a new underlying physical theory beyond the simple model that implements the Higgs mechanism that current searches are based on. This is the sort of thing that keeps theorists like me on our toes as we try to find matter's underlying elements and their interactions. Precise measurements are ultimately what tell us how to move forward in our hypotheses. The Higgs boson is a very special particle indeed and we ultimately want to know as much as we can about it.

Whatever has been found—*the* Higgs boson, the particular implementation of the Higgs mechanism that seems simplest, or something more elaborate—it is almost certainly something very new. The interest from the public and press has been very gratifying, indicating a thirst for knowledge and scientific advances that humanity to a large extent shares. After all, this discovery is part of the story of the universe's evolution as its initial symmetry was broken, particles acquired masses, atoms were formed, structure, and then us. News stories featured members of the public who were fascinated but weren't necessarily quite sure by what. Perhaps the ultimate recognition of the pervasiveness of Higgs boson awareness was the appearance of jokes and spoof news stories indicating the interest—but also some of the bewilderment.

So I'm writing this to respond to many of the questions I've been asked—to share what the discovery means and to explain a bit about where it takes us. Some of what I'll say is already in chapters from my previous books, *Warped Passages* and *Knocking on Heaven's Door*, which provide much of the background to the discovery. These previous books didn't isolate the Higgs boson for extra special attention; rather they covered many topics, including information about the collider, the larger physics story for which this is the capstone, and the nature of science itself. They give the larger context of which this discovery is one—albeit a very important—part. But at least for the time being, the Higgs boson deserves its moment in the limelight. So this essay offers a few new (and old) thoughts. It's an unbelievably exciting moment in physics and I'd like to share some of what occurred and what it means.

The Challenge of Discovery

I guess I was better off on July 4, 2012, than during the last Higgs report in December 2011. On the earlier date, I woke up before five in the morning to do an interview and to listen to talks from CERN, as I was in California and the time zone was not very congenial. At the time of the recent announcement, on the other hand, I found myself on a Greek island where I was taking an all too rare vacation. Although I had poor Internet connectivity and was isolated from my colleagues, at least I was only one time zone away when Joe Incandela, the spokesperson for CMS, first took the stage. Because my somewhat rustic apartment had no Internet, I first learned of the Higgs discovery while sitting in a balcony café—which happily for me opened at 10 A.M., the time of the talks.

In fact, I hadn't imagined when making my holiday plan

that this would happen. I had known the Higgs evidence would increase, but I hadn't known that the engineers would have done such a heroic job in increasing the collision rate, and the experimenters an equally impressive improvement to their analysis methods, that would allow the speakers on July 4 to say with certainty (by physicists' standards) that a particle had been found. One other factor that contributed to the Higgs boson discovery was the decision to run at slightly higher energy—8 TeV rather than the 7 TeV of the previous year—which by itself increased Higgs production by about 30 percent. I was very grateful to the Internet for keeping everyone connected and to Twitter for providing an outlet for my excitement (and for sharing information once people caught on to what was happening and the connection diminished in quality).

Maybe to compensate for that disconnect, I spoke a few days later on a radio program on WNYC. In the pre-show discussion that one typically has before such a program, we reviewed the types of topics that might arise. Most were ones I was prepared for. But I was a little flustered when I was told I would be asked to compete with Dennis Overbye's delightful description of the Higgs boson discovery: "Like Omar Sharif materializing out of the shimmering desert as a man on a camel in *Lawrence of Arabia*, the elusive boson has been coming slowly into view since last winter . . ." (*New York Times*, July 4, 2012).

I certainly wasn't going to think of something as magical in the half-hour I had before the interview—especially as I love that movie. To further complicate the situation, I was on a rock-climbing cliff in Kalymnos, where I had to belay my partner up a climb that we had already set up (I didn't tell the producers that, since they would rightfully have worried about the connection).

So in anticipating my interview, I thought about the question. My partner suggested I say it was the Messiah whom physicists had been awaiting for fifty years, which I thought rather funny but not really helpful. I wanted to create an analogy that reproduced the physics better than as an actor or a deity representative.

What I came up with is not perfect but captures the lead-up to the actual discovery. I said it happened in the way you might find your friend in a crowded stadium full of shouting individuals, where everyone—including your friend—is making noise, but your friend, despite his distinctive voice, is a small peep in the noisy crowd. You would be hard-pressed at first to find him amid the huge din in the background. You might occasionally think you heard his voice, but then it would be drowned out or difficult to distinguish from that of others in the hordes of people.

But imagine now that you knew roughly where to look. You knew what section your friend was in and who he would be hanging out with. So you focus your attention in a particular region. When you hear the sound of his voice, you then begin to be increasingly confident that you have located him correctly. You might not know for sure right away but you might then begin to focus on an even more specific region of the stadium. Eventually you reach the right location where the voice is unmistakable and you know your friend has been found.

The Higgs discovery really did work like that. Higgs

7

events are rare among the far more numerous ordinary particles that are produced. A Higgs boson is connected to elementary particle masses. That means that physicists can predict how it should interact (assuming of course that it exists), so you know its "voice." Alas, the interactions with the ingredients of a proton are in fact rather weak. Quarks and gluons experience the strong nuclear force, which is much more powerful than their interactions with a Higgs boson, so Higgs bosons will be produced only a small fraction of the time and get lost in the "crowd."

Out of the billions of particle collisions that occur every second, only rarely does a Higgs boson get produced. More often than not, collisions result in boring Standard Model collisions that we know to exist. Those collisions give us more detailed information about quarks and the strong nuclear force. But they muddy the waters for experimenters looking for a clear Higgs boson signal.

The only way to find the particle is to have a pretty good idea where to look so that experimenters can distinguish a signal from the "din" of background. *Where* doesn't refer to a physical location like your friend's section in a stadium. Instead it refers to where in the data you expect Higgs evidence to lie—meaning which types of collider events are expected if a Higgs boson exists.

So as with your friend in the stadium, the Higgs boson was initially lost in the background data. Experimenters subsequently looked through trillions of events so that they could begin to see evidence for a bit of a deviation that could signal something special. This was something like the December announcement, where evidence for a particle was

3 sigma level, 3 standard deviations, which is a statistical term meaning the probability is less than one in a few hundred that the result isn't a signal). For those interested in the precise criteria, the level 3 sigma is thought to be too small for discovery, mostly based on experience. Sometimes experimenters don't yet understand all the details of their system or prediction, fluctuations do happen, and everyone knows that if they wait, the answer will sort itself out. That is why experiments require a signal more than 5 sigma, which means that the odds are less than one in 1.7 million that it is simply background noise arising from known familiar particles.

But as time went by and more data accumulated, physicists zoned in on the region that looked a little different and sorted through at least twice as much data. With enough data and enough understanding of the properties expected for a Higgs boson, a clear signal emerged. That signal was what was revealed on July 4 and is almost certainly connected to the Higgs mechanism by which particles acquire their masses.

The Higgs Mechanism, the Higgs Field, and the Higgs Boson

In order to fully appreciate this discovery and what there remains to explore, it helps to know a bit of particle physics, including some deep and subtle underlying concepts. Of critical importance is the distinction between the Higgs mechanism, the Higgs field that is involved in the mechanism, and the Higgs boson particle—which is what an experiment can actually find. Even without experimental proof, physicists were fairly confident about the mechanism, since it was the only consistent way to give elementary particles their masses.

But despite the theoretical consistency of the idea and the failure of any other idea to explain masses, physicists all wanted experimental proof. The experimental results from the LHC have now rather firmly established the relevance of the Higgs mechanism and the Higgs field on which it relies. They have also established the existence of a new particle related to the mechanism—but *the* Higgs boson is part of a very particular implementation, which only further data will definitively confirm or rule out. That is why my title, *Higgs Discovery*, is deliberately ambiguous.

The Higgs mechanism is responsible for elementary particle masses, such as the mass of the electron. Mass is what provides resistance when a force is applied. If particles have no mass, they travel at the speed of light. A particle's mass tells us how it responds to forces and how it travels through space.*

Without its mass, the electron wouldn't bind into an atom, and just about everything else you take for granted about the world wouldn't work either. Such an elementary particle mass, that of the electron in this case, relies on the existence of what particle physicists call a field—a quantity that exists throughout space but doesn't necessarily involve any actual particles. Admittedly, the concept of a field is a

---

*This is a subtle point, but you might want to know that despite the Higgs mechanism's importance, it does not account for most of the mass in the universe. The preponderance of mass in ordinary matter (not dark matter) comes from the mass in the nuclei of atoms and is well understood. That mass arises because of the influence of the strong force, which provides energy when it binds together the three quarks inside protons and neutrons. Due to $E=mc^2$, that energy is equivalent to mass. However, particles that don't experience the strong force, such as the electron, would not have mass without the Higgs mechanism. Neither would the elementary quarks themselves nor the weak gauge bosons that communicate the weak force.

bit esoteric and confusing, especially as the word *field* outside of physics conjures images of cows grazing, which became clearer to me when I read the word *champs* in French physics textbooks.

But really we encounter fields (in the physics sense) in many contexts. A magnetic field is perhaps the most familiar one. When you hold a magnet close to your refrigerator door, you feel a force. There is "nothing" between the magnet and the refrigerator, which is to say there is no actual matter there. But there is a field and that field is responsible for the tug you feel that attracts the magnet to the shiny white surface.

That field is of course local. You move a little away from your kitchen and the influence of the magnet is too negligible to feel. The Higgs field, on the other hand, is everywhere. It is spread throughout the universe. The field isn't made up of actual particles. In a sense it involves something like a type of charge spread everywhere through empty space. Particles that experience the weak force (which is to say leptons like the electron, quarks, the weak gauge bosons that communicate the weak nuclear force, and the Higgs boson itself, as we will soon see) interact with that "Higgs charge" and thereby acquire mass. In the presence of the Higgs field, particles have masses. The heaviest particles interact with the field the most, and the lightest the least.

The Higgs boson, on the other hand, is an actual particle—a fundamental object that has definite mass and interactions. Although not a field itself, the particle is indeed associated with the Higgs field. Essentially when you jiggle the Higgs field—add a bit of energy—you can create an actual particle. A single field both permeates the vacuum—empty space—with a nonzero constant value everywhere, and is also responsible for particle creation.

From the point of view of particle masses acquired through the Higgs mechanism, the Higgs boson is an appendage that comes along for the ride. The Higgs boson itself is not essential to particle masses. Clearly masses didn't just come into being on July 4, 2012, with the announcement of discovery. The Higgs field—not the Higgs boson—gives masses.

Yet the Higgs boson is the telltale sign that the theory of the Higgs mechanism is realized in nature. With enough energy, a Higgs field can create these particles, and they have properties specific to the role of the Higgs field in allowing for particle masses. On July 4, we learned that the Large Hadron Collider at CERN had finally made enough of these Higgs bosons to give physicists a clear signal that couldn't be mimicked by Standard Model particles we already knew to exist. In other words, the signal could make sense only with a new particle present.

The Higgs boson—or a Higgs boson surrogate whose properties deviate from that of the simplest implementation of the Higgs mechanism—is also essential to our learning how the Higgs mechanism actually came about. Future data are guaranteed to offer more details about the particle's properties so that we can definitively establish whether it is part of a simple sector of particles implementing the Higgs mechanism, or an even richer one, involving more Higgs-like particles or additional structure involving new forces and interactions.

It's interesting to reflect on the implication of this discovery for our understanding of "empty" space. When we describe space as empty, we generally mean that no actual matter is present. Matter is stuff that clumps together under the force of gravity to form structure. The Higgs field is not like that. It takes a nonzero value but remains uniformly spread everywhere.

The Higgs field is not composed of actual matter. In a sense the Higgs field carries a type of charge and furthermore allows that charge to appear and disappear into it. We don't directly feel that charge (associated with the weak nuclear force) because the force associated with it has such short range. It does, however, allow for nuclear beta decay, for example, by which a neutron can decay into a proton, an electron, and a neutral particle called a neutrino.*

Particles have interactions directly with the Higgs field that don't involve the Standard Model forces at all. In a sense, the Higgs boson itself communicates a kind of force, but one very different from those we know about involving only a fixed unit of charge. Particles' interactions with the Higgs field allow for a wide range of masses because each particle has its own individual interaction with the field.

You can ask whether the Higgs field carries energy too. We don't know the answer, since all that can be measured is the gravitational influence of the net energy of all the fields in the universe. This energy takes a nonzero value known as the dark energy, which is a rich and fascinating subject in its own right. The dark energy is also associated with empty

*More precisely, the decay product is an antineutrino, the antiparticle of the associated neutrino.

space—it is the energy empty space possesses. Einstein taught us that any source of energy has consequences for gravity, including the absolute energy of an empty universe. The energy carried by empty space has measurable consequences, such as the acceleration of the expansion of the universe. The fact is, empty space is not truly empty. It can have energy and charge. It just doesn't have matter.

Even though we don't know the energy carried by the Higgs field, we do know that the energy depends on the value the Higgs field takes. The energy is lowest when the Higgs field is nonzero. The field is a bit like a pencil standing on end: the zero value is analogous to the pencil standing upright, whereas the nonzero value, which carries lower energy, is analogous to the fallen pencil lying on its side. The Higgs field too doesn't want to be at the symmetric point with zero value but prefers to "fall" in some direction where it breaks a symmetry. When it does so, it takes on a nonzero value that is ultimately responsible for elementary particle masses. Admittedly this is all a bit abstract in nature, since it doesn't involve matter we can throw around and touch. But one of the beautiful aspects of the Higgs mechanism is that it tells us about the richness of empty space.

Before ending this section, I'll answer one more interesting question that I've been asked. Where does the mass of the Higgs boson itself come from? The answer is that the Higgs boson interacts with the Higgs field. So just as with other elementary particles, the Higgs field accounts for the Higgs boson's mass.

You can think of a physical interaction as two particles entering a collision, and then two particles emerging, most likely in another direction. That is what happens when an electron scatters off another electron, for example. In the case of the Higgs boson, the same interaction that permits two Higgs bosons to collide and merge also allows a single Higgs boson to interact with a background field—the field existing in empty space. This is key to understanding one of the questions physicists ask as we move forward: What does the mass tell us about the Higgs' interactions with itself? This self-interaction, as it is known, might be determined by physics that goes beyond the Standard Model and is one more important clue—in addition to the different Higgs boson decay rates—of the nature of the particle and what lies beyond.

Higgs Boson Decays

A key property of the Higgs boson—one that is essential to understanding how it is found—is that it is extremely unstable. It lasts only a fraction of a second before turning into Standard Model particles such as quarks and leptons. "Decay" means that the Higgs boson itself ceases to exist and instead decay products (familiar Standard Model particles) are created and carry away its initial energy and momentum.

That means that when experimenters search for a Higgs boson, they don't look for the particle itself but for the particles into which it decays. By adding up the charge, energy, and momentum of those final state particles, they can determine if their origin was a particle with definite charge (zero in this case) and a particular mass.

If a real particle decays, there is a real mass involved. So when you plot the number of events you have versus putative mass, you would find a bump—an excess of events that is centered at the actual mass of the physical particle.

That is what happened with the Higgs discovery, which is why you might have heard experimenters discussing their discovery in a language using the word "bump".

The bump isn't just a line, however, which is what would happen in a world where all measurements are perfect. Because they are not, the measurements are centered on the physical value but there is a smooth drop-off away from the exact Higgs boson mass—a drop-off characteristic of slight mismeasurements. With increased data, the measurement improves and the signal becomes a narrower peak.

In fact, even if measurements were perfect, there would still be a spread of values, but a much smaller one—because of quantum mechanics. The Higgs boson decays; it doesn't last for ever. The quantum mechanical uncertainty relation permits the Higgs boson mass to look wrong for a short amount of time (less than its lifetime). Those Higgs bosons with slightly wrong masses get recorded to generate what is known as the Higgs width. That width is in fact a measure of the Higgs boson lifetime. Unfortunately, at the LHC, it is too small to measure, since the larger experimental uncertainties give more uncertainty to the mass that is measured.

Actually, as if experimenters didn't have enough on their plates trying to find the Higgs needle in a haystack, there is one more confounding problem that makes measurements even more challenging. This problem is known as pileup. The LHC is a remarkable machine for at least two reasons.*
One is that it has higher energy than any accelerator that

---

*Much more about the LHC and the CMS and ATLAS experiments can be found in *Knocking on Heaven's Door*.

existed before. The other is that it has a remarkably high intensity. That is, the rate of events is enormous. In fact, it is so enormous that more than two protons collide when a bunch of protons within the beam collide.

At full intensity, the number of protons is enhanced to the point where about thirty collisions occur at virtually the same time. Most are uninteresting and don't confuse the data too much. But when doing a precision study such as a Higgs boson decay, you can't be too careful, and the experimenters employed a number of clever techniques to identify which particles are really part of the collision of interest. That involves identifying the right decay products that came from the Higgs boson decay. The next section is a more nitty-gritty one that tells what those might be.

Decay Modes

A Higgs boson decays, converting into other particles that carry away the particle's initial energy, momentum, and charge. Detectors such as those in ATLAS and CMS measure all those properties of the decay products which fly away from the region where the protons initially collided. Experimenters have to add up the energy and momentum of all the particles emanating from the collision region where the Higgs boson is produced and decays to figure out the properties of the particle that was momentarily in existence there.

The connection between the interactions of the Higgs boson particle and Higgs boson field allow physicists to predict the Higgs boson's interactions based on particle masses. Those interactions are important to studying the Higgs boson because they are what permit the Higgs boson to decay.

The particles into which the Higgs boson decays also have

to have a sufficiently large interaction with the Higgs boson that such a process can occur. If a Higgs boson didn't interact with a particle, it couldn't decay into it. The Higgs boson is a very special particle whose interactions are connected to particle masses: heavier particles interact more and lighter ones interact less. So a Higgs boson decays the most into the heaviest particle that is not so heavy that energy won't be conserved in the decay. The bigger the mass, the bigger the interaction.

But if the mass of the particle is too big, there won't be enough energy to make it. The particles into which the Higgs boson decays must be sufficiently light to allow energy to be conserved in the decay—along with momentum and charge, which the decay products all carry away.

The bottom quark is the heaviest particle for which twice the mass is still less than the measured mass of the Higgs boson—125 GeV—so that a bottom quark and its antiparticle, the bottom antiquark, can be produced. Because it is the heaviest particle for which the decay can occur, most Higgs boson decays are into bottom quarks and antiquarks.

To deconstruct that paragraph: GeV is a funny unit of mass that particle physicists use. It is actually a measure of energy. GeV means giga electron volts, which is a billion electron volts. We also sometimes speak of TeV (tera electron volts), which is the equivalent of 1,000 GeV. The collider currently runs with 8 TeV of energy. Energy can be used as a measure of mass too. Einstein's famous formula $E=mc^2$ tells us that energy (E) and mass (m) can be used interchangeably, since c, the speed of light, is constant.

The bottom quark is a type of elementary particle. You might have heard of the up and down quarks that sit inside a proton and a neutron. That is to say, the protons and neutrons sitting inside the nucleus of an atom are not elementary but are in turn composed of more fundamental particles called quarks. Those quarks are held together by the strong nuclear force—yes, that same force I referred to earlier that often produces Standard Model particles when protons collide. They are the two lightest quarks. One has positive charge (the up quark with $+\frac{2}{3}$ charge), and one has negative (the down quark with charge $-\frac{1}{3}$). Two up quarks and a down quark together give the proton its charge of $+1$.

But those aren't the only quarks. One of the biggest mysteries in particle physics is that for every familiar type of particle (by familiar I mean ones that exist on Earth today), there are heavier versions that have bigger mass, are unstable, and are created on Earth only in accelerators and, for some, in cosmic rays.

The bottom quark is one of those heavier quarks. Like the down quark, it is negatively charged and in fact is the heaviest of the three quarks that carry that same charge (the other is the strange quark). The reason the Higgs boson decays primarily into the bottom quark is that it has this relatively heavy mass and therefore interacts more with a Higgs boson, whose interactions are determined by particle masses.

But unlike a bottom quark, a Higgs boson has zero charge. If charge is to be preserved, the Higgs boson can only decay in a way for which the net charge of the decay produced—the sum of the charges of all the particles into which it decays— is zero. An antiquark carries precisely the opposite charges of a quark. And a bottom antiquark carries precisely the opposite charge of a bottom quark. The charges of the bottom quark and its antiquark add to zero, which is the charge of the Higgs boson that decayed into them.

That is good because in addition to electric charge, a bottom quark carries another type of charge related to the strong force. A bottom quark is not neutral under either the electric force or the strong nuclear force. However, a bottom quark and a bottom antiquark together are neutral. A bottom quark and antiquark together can carry no net charge whatsoever. That is exactly the property of a Higgs boson—it carries no net charge. A Higgs boson does interact directly with quarks and other elementary particles that have masses. But those interactions are not merely through the three known Standard Model forces (electromagnetism, the weak nuclear force, and the strong nuclear force; I'm leaving out gravity since it's so extraordinarily weak). Direct interactions related to those that are responsible for masses always occur.

So a Higgs boson can decay into a bottom quark and a bottom antiquark without violating any known conservation law. And it does just that. Of course, the Higgs boson interacts with heavier particles too—the weak gauge bosons and top quark in particular. But a 125-GeV Higgs boson isn't heavy enough to decay into those particles. The sum of a top quark and antiquark mass is far in excess of 125 GeV. The sum of two charged weak gauge boson masses is about 160 GeV, which is again too heavy for a Higgs boson to decay directly into those particles.

So the dominant Higgs boson decay is into bottom

quarks. But here's the rub. At the LHC, bottom quarks are tough to identify and distinguish from the Standard Model "din" (for physicists, known as background). They are quarks, and so many quarks get produced at the LHC that you need to look at some special production mode in order to isolate the Higgs decay signal.

When the Higgs boson announcement was made, the two modes that dominated the signal were decays into photons and decays through weak gauge bosons—not the more frequent decays into bottom quarks.

I realize this is a lot of information, but if you're a little perplexed at this point, you should be. I said the Higgs boson interactions correspond to elementary particle masses so that it should interact more with heavier particles. The photon has zero mass, which makes it seem as if the Higgs boson shouldn't interact with photons at all. On top of that, I just said the Higgs boson is too light to decay into weak gauge bosons. So what am I talking about?

I'm talking about quantum mechanical effects. The most visible decays of the Higgs boson to date occur because of processes that are allowed only when we take quantum mechanics into account. The Higgs boson can in fact decay into two photons, but only because quantum mechanics permits it. When the CMS and ATLAS experiments were designed, a strong consideration was being able to measure the energy and direction of photons as well as being possible to allow for a precision Higgs boson discovery.

Quantum mechanics allows a Higgs to decay into two photons because it allows for virtual particles—particles that don't exist for ever and in fact don't even have the correct mass to be the actual particle that can survive in the real world. Virtual particles are a big deal and one of the many nonintuitive properties that quantum mechanics permits. They are particles that briefly come into existence, can have further interactions, and then must disappear.

Now, even though a Higgs boson wouldn't interact with photons according to classical physical laws (those that don't take quantum mechanics into account), it will interact with them according to quantum mechanical laws. What happens is a Higgs boson turns into particles too heavy to be made in nature, such as weak gauge bosons and top quarks—the heaviest of the Standard Model quarks. The heavy particles disappear, annihilating each other, but in doing so emit photons. Even though the interaction between a Higgs and two photons is suppressed (all interactions permitted only through quantum mechanical processes are), the overall rate for decay into photons is small but not negligible. The Higgs boson decays into photons about a fifth of a percent of the time. And this is sufficiently often to generate a clear photon signal once enough Higgs bosons have been produced.

Let's now return to the decays involving weak gauge bosons—the other decay that was important to discovery. Those in fact happen a little differently. I wish this were all more straightforward to explain, but I'm giving you the real story here. Weak gauge bosons also get produced as virtual particles—those that have the wrong mass to be the actual physical particle that survives in the real world. But they can have almost the right mass. After all, half of 125 GeV—the amount of energy each weak gauge boson has at its disposal—isn't that far from 80 GeV—the mass of a charged

W boson, or from 91 GeV—the mass of a neutral Z boson. It's sufficiently far away that two real gauge bosons cannot be produced. But it's sufficiently close that the rate of decay through two weak gauge bosons, after which the weak gauge bosons themselves decay, is in between that of a classical and a quantum mechanical process. So for this particular mass of a Higgs, decays through weak gauge bosons occur reasonably often as well.

Experimenters look for these decays by searching for the decays of two gauge bosons. These are not just any two gauge bosons, however. They are gauge bosons that emerged from the decay of a Higgs. This means that when you add up the energy and momentum of the decay products and put it all together, it will reproduce the mass of a Higgs boson. So these events are distinguished from other events by contributing to a Higgs boson bump when you plot the masses that experimenters put together from the decays.

The Development

So now that we know a bit more about what experimenters actually look for, let's return to how the discovery played out over the course of the past seven months.

The LHC has now been running for about two and a half years. One of the chief search targets all along has certainly been the Higgs boson. The interesting thing about the Higgs boson was that combined experimental and theoretical considerations told us to expect it to be relatively light, which is to say well within the kinematic reach of an LHC that was running at only half energy. We didn't know for sure this would prove correct, but it was promising.

However, because the particles that create the Higgs boson are light (and therefore have small interactions with it), and because the Higgs boson can be best observed in modes of decay that happen only infrequently, experimenters needed many collisions before they could see a signal that sufficiently dominated over background. So everyone had to wait as data were collected.

Of course it was worth it. The first real hint that something was afoot (at least to those of us not on the experiments) came in December 2011. Even in my sleep-deprived state many time zones away in California, it was clear that something exciting was occurring. There was evidence that a particle was decaying to two photons at a rate that exceeded the Standard Model expectation in the absence of a Higgs boson. ATLAS especially had something that looked like a signal, while CMS had data that certainly didn't rule out that possibility but on its own might not have looked as strong. The CMS data improved over the next few months, however, so both experiments had evidence of a signal of something new—just not strong enough to claim discovery.

The reason physicists require a big signal to claim discovery is that the search requires evidence of an excess of events over background—the "bump" I referred to earlier. But no one can predict exactly what happens in any particular collisions. The predictions apply to the average. With only a few collisions, the data might contain a spurious signal—a fluctuation that exceeds background expectations and accidentally mimics a Higgs boson. Only with enough data—enough collisions—will the probability that you are seeing just a fluctuation become sufficiently unlikely that you can really believe you have discovered a new particle.

Personally, I was gratified by the December result. It was the best we could have hoped for at the time unless our predictions were wildly wrong. There simply hadn't yet been enough collisions to make a true discovery so it is what

I expected when I heard hints that there could be a signal. In fact, I'm on record as saying this since I answered Dennis Overbye's questions for the *New York Times* and did a radio interview immediately before the announcement—a bit of a risk since, as a theorist who isn't a member of any experiment, I didn't have much more information than anyone else until the result then was made public.

The subsequent months were very exciting for physicists. Theorists like me had to take into account limits set by the LHC and the possible existence of a 125-GeV Higgs boson, and I and many others developed models that did so. In January I visited CERN, the major physics center near Geneva that houses the LHC, and had valuable conversations with physicists there. Experimenters shared their insights and were eager to learn more about how to search for models that I was working on at the time.

In March I had yet more contact with the experimental community at a major international conference named Moriond, that delightfully takes place at an Italian ski resort (which suffered, I'm afraid, from prematurely warm weather). People there were still discussing "faster-than-light neutrinos" measured by the OPERA experiment, which are definitely excluded now. But they had only speculation and some modest measurement improvements to share about the Higgs boson. The physicist Greg Landsberg expressed his frustration at needing to talk about Higgs results after drinking "phantom of the opera" grappa. There really was no discovery to report, and the updates were not overly exciting in March.

In fact, on asking around, I found that virtually no one

29

thought a discovery would be announced by the time it was. Discovery arrived more swiftly because engineers led by Steve Myers managed to crank up the intensity of the machine and experimenters significantly advanced their analysis techniques.

As for whether people believed the Higgs result of December was real, the majority expected it to be confirmed. However, you might be surprised to learn that many theorists had hoped the signal would prove incorrect because of the deep consequences it would have for the underlying theoretical construct. If there were no Higgs boson, consistency would have required something even more surprising and interesting.

I have to say, given my dealings with the public—and with experimenters—I found it hard to share in that hope, as interesting as it would be. Experimenters deserved to find something. They had worked long and hard—and waited a great deal too. Setting limits is important, but a discovery is something else altogether. Finally they have a particle to measure and understand more about.

And, I must confess, as challenging as it is to explain why a Higgs boson discovery is interesting, explaining why not finding it would be more interesting was a task I was glad to avoid.

One of the funny things about the Moriond conference, however, was that even though so much attention was focused on the Higgs boson, most talk titles used the words "scalar particle" or "scalar boson" rather than "Higgs boson." I later found out that this was because François Englert, who was one of the original six physicists—Peter Higgs, Robert

Brout, and François Englert, with the collaboration of Gerald Guralnik, C. R. Hagen, and Tom Kibble—who developed the so-called Higgs mechanism, was in attendance as well. I will not enter into a discussion of priority claims, but the mechanism was the capstone to an edifice that has been in the making since Marie Curie first discovered radioactivity, showing that more happened at the subatomic level than anyone had imagined, and culminating —as of today—in the Standard Model and the Higgs mechanism.

Englert gave a wonderful talk about the particle and the mechanism, and I was surprised by how engaged I found myself, even though I was familiar with the material. One of the true delights at Moriond was getting to know François Englert a bit better. Peter Higgs is an interesting character too, but I don't know him myself. When I visited Bristol for the Ideas Festival there, I learned that Peter Higgs's interest in physics was sparked in part in Bristol, where he attended the same school as Paul Dirac (the physicist who developed the idea of antimatter).

Higgs had originally planned to study engineering—as did Dirac for that matter. Having then decided to study particle physics, he was initially dissuaded and studied molecular physics before happily returning to the type of physics that truly interested him and making his major breakthrough almost fifty years ago in Edinburgh.

The Belgian François Englert had perhaps an even more surprising story. He too had studied engineering and had been sent to England "to work on cables," as he described it. Apparently his major accomplishment there was to incite a strike, for which he was rewarded with pay on condition that he leave right away. Having had little interest in cables in the first place, Englert told me this was an offer he could not refuse. He returned home and started to work on semiconductors, which was closer to the physics he would ultimately study, but he was working in a laboratory and not doing the type of theoretical work that would continue to hold his interest.

Englert's next step was the Belgian army. Fortunately for him, it had a good library and he managed to find an adviser in condensed matter physics and get a degree while doing his required service. When he moved to Cornell for a postdoctoral fellowship, he had the good fortune to work with Robert Brout, who would become his major collaborator. The collaboration was so successful that Cornell offered Englert a job, which he refused in favor of returning to Brussels, where he did not yet have employment. And more remarkably still, Brout—though tenured at Brussels and not yet employed at Cornell either— decided to join him. In any case, the story continued remarkably satisfactorily. They both got jobs, and they went on to do work for which Englert is likely to win a Nobel Prize (sadly, Brout recently passed away). They played a significant—and daring—role in Belgium, where the new ways of thinking about elementary particle physics had not yet caught on.

I also had the satisfaction of introducing François to a young French professor who worked on the Higgs boson search and watching Englert's interest and excitement at the then current experimental situation as they discussed the physics into the night. Clearly the moment of truth was approaching for Higgs, Englert, and the other potential laureates involved in the Higgs theory development, as well as for the rest of us.

So what happened between March and July? Evidently quite a bit. The engineers and experimenters were hard at work. Theorists were too, but we were not privy to the daily updates in data, since experimenters don't want false rumors leaking out. So I personally did not know much more about what was to come. I knew the Higgs result would improve because someone had told me that a lead experimentalist had smiled when viewing the data (yes, that is the kind of tea-leaf reading we are sometimes reduced to), but I didn't know a discovery would be announced.

In fact, I was not alone. At a European conference I attended a week after the announcement, I had the opportunity to talk to Rolf Heuer, the director general of CERN, and Steve Myers, the chief engineer. Both told me of the uncertainty—and excitement—of the days and weeks preceding the announcement. Every day the results were different—though converging. When the CERN seminars were scheduled only two weeks before July 4, no one was yet certain of the results. In fact, even up until the last week and the last few days, the numbers were still fluctuating. Claiming a discovery is a very big deal, one that everyone involved took very seriously. It was only in the last days that it became clear that the word "discovery" could be used.

Of course, by July, members of the experiments knew what was happening. At CERN, many others certainly did know something was afoot by the time of the announcement. Students and others who were not sufficiently elevated in the CERN hierarchy even camped out the night before to guarantee themselves a seat in the auditorium—which turned out to have been worth it when, in the morning, hundreds of people were turned away.

Joe Incandela, the spokesperson for CMS, gave the first talk. He began by discussing the modes with the strongest signal, the decay into two photons and the decay into neutral weak gauge bosons. After calmly discussing the results, he put up a slide saying that the two modes together gave the 5 sigma signal that had come to be the physicists' benchmark heralding discovery. It was extraordinary. The CERN audience broke into applause when those words were spoken. In my outdoor café, I could do no more than tweet to vent my excitement.

The ATLAS talk followed and their results were just as exciting—even if the Comic Sans font designed for children that was used in the presentation was the source of great amusement on the Internet and among my tech-savvy friends. After Joe Incandela's talk, Fabiola Gianotti, the spokesperson of ATLAS, spoke about their 5 sigma and was applauded too. In fact, she had to remind the audience that her talk wasn't yet over so that she could finish.

The remarkable conclusion of the seminars was that both CMS and ATLAS had made a discovery and it was very likely connected to the Higgs mechanism. Rolf Heuer happily summarized the situation after the two talks. We don't yet know for sure whether it is the simple sort of Higgs boson first proposed or something more complicated that plays the same role, but we do know that a particle connected to the Higgs mechanism has been found.

In fact, both experiments have been lucky (or there is something very important yet to be understood) in that the signal in two photons seems to fluctuate upward from what is expected in the Standard Model (meaning there was a slightly larger bump than was expected). For CMS, the photon channel is high but the neutral gauge boson signal is low. So are signals in a couple of other types of decays that ATLAS hasn't even checked yet. In ATLAS, not only did the photon data exceed expectations, they were lucky with the neutral gauge boson decay mode too.

Gianotti pointed out during her talk that we are very fortunate with the Higgs boson mass. There are in fact five different decay modes that occur often enough to be studied in detail. Had the Higgs boson been heavier, decays to weak gauge bosons or top quarks would have overwhelmingly dominated. Had the mass been any lighter, the decays through virtual weak gauge bosons would have constituted an insignificant fraction of the total decays. But for this particular mass, essentially all measurable decay modes are rare. No single one dominates the data. Nonetheless, all occur with enough strength that future LHC-running will permit detailed investigation. This is critically important in establishing whether the discovery was of a single particle called the Higgs boson or was part of a bigger story in which Higgs boson properties aren't exactly those predicted in the simplest model.

I'll take this opportunity to make a brief digression about particle accelerators. Before the LHC was constructed in Europe, Americans had built the Tevatron, a smaller accelerator that collided protons and antiprotons at one-quarter the current energy of the LHC (and one-seventh the LHC's ultimate energy). The energy and collision rate were not sufficient to discover a Higgs boson at the level of rigorous standards we particle physicists require. But they certainly could have measured Higgs boson decays. Furthermore, this machine was a proton–antiproton collider, which made a particular Higgs production mode much more important. So unlike the LHC, the best Higgs signal was in bottom quarks.

So although today we celebrate the LHC's discovery, my hat is off to the Tevatron experimenters too. They did in fact measure an excess of events involving bottom quarks that seemed to be centered at the right mass to correspond to the Higgs boson, but much less than required for a discovery. Even so, Heuer remarked to me the reassurance that came from the supporting evidence the Tevatron provided.

Although the Tevatron has now ceased to operate, the LHC will provide lots more information about the different Higgs boson decay modes, including decays into bottom quarks, photons, weak gauge bosons, and tau leptons (heavier versions of the electron). Further measurements of the different decay modes should determine whether or not the slight deviations from Standard Model expectations are real reflections of the physical world or mere statistical fluctuations in data that will disappear over time. This will require more collisions.

CERN's plan was to shut down the LHC for a year and a half at the end of the year, during which time the machine will be refurbished and upgraded. When the machine turns back on, it will have considerably higher energy. This higher energy will give the LHC a huge boost in the search for new particles and new physical theories.

The good news that Rolf Heuer shared within a week of the Higgs announcement is that CERN will keep running the LHC a few months beyond the previously planned shutdown. With this increased running time, experimenters should be able to collect enough data to make the necessary determinations on the nature of what has been found. A few months might not sound like much, but the increase in the collider's intensity means that later data carry more weight than earlier data. With enough collisions, the properties of the Higgs boson will be much better known, and that is good news for those of us who want to know in which direction to head during those eighteen months while the collider is dormant. Either way, a large community of physicists and others look forward to the result.

So new data might soon tell us if the particle that was found is *the* Higgs boson or if it provides hints of something beyond the Standard Model. It might also tell us more directly about other new particles and interactions, though realistically that might require at least the higher energy (13–14 TeV) that the LHC will have after the eighteen-month shutdown and upgrade.

Whatever is found, physics will move forward. We will either know that the simple Higgs boson has been found, or we'll have evidence of physics beyond the Standard Model that will point the way forward. These detailed studies are very well motivated. The Higgs boson discovery is more likely to be the beginning of the story than an end.

A Fundamental Scalar?

Now that a Higgs particle has been found, of course the big question is where to go from here. Yes we can take a few moments to celebrate and drink champagne, as LHC experimenters tend to do. But ultimately physicists want to put together a bigger picture.

I've discussed the existence of a Higgs field, which is a quantity that permeates the vacuum and has a nonzero value that is the root of elementary particle masses. A nonzero value for a field is indeed something special. If a field carries charge, it means that charge can disappear into the vacuum, so the charge won't be conserved. If the field changes under rotations, much as an arrow pointing in a particular direction would do, the vacuum wouldn't preserve rotational symmetry. And if the particle that is created by the field has nonzero spin, the rotational symmetry (and Lorentz symmetry, Einstein's extension of rotational symmetry that includes time) would be broken as well.

This means that the particle created by the field also has to be special. It has to have zero spin, which is to say it must be a scalar particle, which is a particle with spin 0. Spin is the property of a particle that tells us how we can expect it to behave under rotation. Particles like the photon have spin 1. Particles like the electron or a quark have spin ½.

For particle physicists, spin 0 is equivalent to saying that the Higgs particle (created by an associated field) doesn't change under spacetime symmetries such as rotations. An electric field turns on only when actual charged matter is present. Otherwise, rotational symmetry would be broken. The Higgs field with zero spin can turn on in the absence of any particles whatsoever, since no symmetry (such as rotational symmetry) is broken when it does. The Higgs field is the same in all directions.

The reason this is so potentially interesting is that up until now no one has discovered an elementary scalar particle. Scalar particles do exist, but the ones people have observed so far are composites made up of more fundamental particles such as a quark and an antiquark.

The Higgs boson has the potential to be the first elementary scalar particle ever found. This means that the Higgs boson in some sense is a truly new form of particle. You can even think of particles that interact with it experiencing a new type of force, one distinct from the four known forces.

But there is a mysterious side to a fundamental scalar. According to calculations based on quantum field theory, which combines quantum mechanics and special relativity, fundamental scalars should be extremely heavy—sixteen orders of magnitude heavier than the boson that has been measured. Such calculations indicate that the Higgs boson should be enormously heavier than we know it to be.

Even before the LHC measurement, physicists knew the mass should not be very different from the masses of the weak gauge bosons—near an energy associated with a symmetry breaking, which is about 250 GeV. I want to be clear that this is an approximate criterion. We don't necessarily require a precise 250-GeV mass; 125 GeV is just fine. But we don't want the mass prediction to be 10 million trillion GeV. Nonetheless, without additional underlying physics, a light fundamental scalar is an enormous fudge, or what we call "fine-tuning."

This is the hierarchy problem of particle physics (discussed further in *Warped Passages*). The answer to this conundrum is likely to involve some deep new under-standing of nature, which could be a new symmetry of space and time, such as the theory known as supersymmetry, or even an extension of space itself, such as a warped extra dimension.

In any case, even if *the* Higgs boson exists, it is most likely part of a larger sector of new particles. That would be a big story beyond the Higgs that we hope to learn more about next from the LHC.

Now That We've Found Higgs, What Do We Do With It?

We will all learn more as the LHC continues to run. With further data, we'll learn if indeed *the* Higgs boson has been discovered. We will also have further constraints on new physical theories as searches continue if they don't find anything, and of course if they do find something else new, it will point the way forward.

Really, although most people hesitate to claim the particle that was discovered is *the* Higgs boson—a particle in a specific model that implements the Higgs mechanism responsible for elementary particle masses—the discovery was made because the properties look an awful lot like those we would expect from this particular particle. At the level of accuracy of the measurements, the data conform to Higgs boson expectations, giving few hints so far as to what lies beyond the Standard Model. Even if it turns out to be part of a more complicated model than the one that gives rise to a single Higgs boson and nothing else, whatever is found is part of a theory associated with the Higgs mechanism yielding elementary particle masses.

However, until all the different decays are measured with better precision, it still isn't absolutely clear whether a simple Higgs boson of the sort that seems to have been found was responsible for the Higgs mechanism or something more complex. But even what we know so far severely constrains the possibilities that theorists like myself can now think about.

On top of that, the Higgs boson, even if it is *the* Higgs boson, is almost certainly not the only particle yet to be discovered. The LHC was not designed simply to look for a single particle. It is searching for yet richer elements that can underlie the Standard Model. Knowing that the Higgs boson is part of the sector associated with the Higgs mechanism certainly addresses one mystery—how elementary particles acquire their mass—but still leaves open the question of why those masses are what they are.

Forging ahead, there are several agendas both for experimenters and for theorists. Experimenters will have to measure this particle's properties with much better precision. That will allow us to determine with greater certainty what in fact has been found. They will also be searching for extensions of the Standard Model that go far beyond this one additional particle. Any extension will almost certainly involve a sector with many additional particles and hopefully many additional experimental signals.

Theorists on the other hand will continue to pursue the hierarchy problem. But we will also puzzle over the particular value of the mass that now seems to be established. Some of my recent research involved trying to reconcile what the LHC has measured with proposed theories that have already been suggested. New data might quickly obviate any such speculations. But they give a flavor of the kind of work we do. And of course if we're really lucky, they could turn out to be right.

My investigations have focused in part on supersymmetry. If the world is indeed supersymmetric, for every known particle there is a partner antiparticle that has the same charges but different spin. The superpartners, as they are known, are expected to be heavier than the known Standard Model particles. They are expected to be light enough, however, that the LHC will have sufficient energy to produce them.

The problem is that current experimental limits are beginning to make ordinary vanilla supersymmetric models look increasingly unnatural. On top of that, the value of the Higgs mass stretches this same vanilla supersymmetry to its limits. The theory is motivated in part by justifying a light Higgs boson and somehow just doesn't want the Higgs boson to be this heavy. The simplest models predict that it's much lighter. The question becomes, does this mean that theorists who explore supersymmetry have been entirely on the wrong track, or have the models that have been proposed been too simplistic? Certainly, before throwing out such a beautiful idea, we would want to explore all the options.

Almost by accident, my collaborators Csaba Csaki and John Terning and I stumbled upon a possible explanation for both these phenomena. In our version of super-symmetry, some superpartners have big masses, whereas others do not. This turns out to be an important difference for LHC searches, since the particles that should be light are not the ones that have been well studied (at least so far).

On top of this, additional Higgs boson interactions automatically allow for a bigger Higgs mass. In fact, when we first started working on our paper, we were focused on the theoretically interesting aspect of two different scales for new particle masses. When we were finishing our paper in December, it was almost an afterthought to mention that our model readily accommodates the Higgs boson mass that was first suggested at that time. At the last minute, we remembered to change our abstract. With the Higgs mass confirmation, I'm glad we remembered.

In general, the search strategies for a modified version of a supersymmetric model—or other extensions even to known ideas—are likely to be very different. One of the chief reasons model building in particle physics is so important is to ensure that all possible signatures of interest are searched for. The number of collisions at the LHC is so enormous that unless experimenters have some definite target in mind, the process of interest can be buried in the data.

But the LHC will explore other avenues too. Dark matter searches are on the agenda, as are searches for new heavier particles of various sorts. The Higgs boson discovery is inspiring, in that the particle was predicted and found. But most of us are humble enough to realize that nature can have surprises in store.

Religion, Utility, and All That . . .

Since I'm addressing some of the questions I've been asked since the discovery, I'll take a moment to address three that I've heard a lot: What is this useful for? What does it tell us about religion? What do you think of the name "God particle"?

You might guess that I'm not a big fan of the name "God particle." One interviewer protested that the particle is very important and is critical to matter as we know it. Perhaps out of boredom (or maybe because I was in Greece), I pointed out that many particles are critical to matter as we know it. In a monotheistic universe it would be an overstatement to single out the Higgs boson as a deity. Maybe in a pantheistic universe, we could have many god particles. But really, they are particles and have nothing to do with religion. We build up matter at different scales, as I describe in *Knocking on Heaven's Door*. The Higgs boson is important but it has a very well defined role.

On a related topic, the Higgs boson discovery says nothing about religion. Surprisingly, several interviewers thought it did and that it would even bring people to church. I would think this discovery would lend credence to the scientific method, and perhaps spark curiosity about understanding the world through science. Even those of us who trust the scientific method are very excited when a prediction proves true. After all, what the scientific method allows us to do is both rule out and verify theories by testing their experimental consequences. The Higgs boson prediction turned out to be right. This was based on theoretical considerations that took into account existing measurements. It is a tribute to science and the ingenuity of both theorists and experimenters that such a prediction could be made and verified. The discovery is truly inspirational—in a scientific way.

And is it useful? The Higgs boson has no practical implications that we know of. But believe it or not, no one knew what the electron was good for when it was first discovered. The same applies to quantum mechanics, which was critical to semiconductors and the current electronics industry. So not being able to think of practical applications isn't overly surprising.

We do know that the discovery is good for piquing humanity's curiosity and rewarding our ability to ask—and answer—deep and fundamental questions. Societies accompany advanced science with advanced education and generally with a thriving economy that derives directly and indirectly from scientific developments. After all, powerful computing methods, important magnet developments, and precision electronics all are needed to make the LHC and its experiments work. Superconducting magnet technology that was developed for accelerators is now used in medical and industrial applications. And the World Wide Web was developed at CERN to allow efficient information transfer among collaborators in different countries. Technical engineering advances—along with the mathematical and theoretical advances that inspire students and the public—help advance societies.

But really, scientists have discovered a new particle—one that tells us about the power of empty space. It was predicted almost fifty years ago based on theoretical considerations and the need to make the Standard Model consistent. It was verified through heroic engineering and experimental techniques. The particle's discovery is tremendously exciting. It's also inspirational. Let's just enjoy that for now.

For more in-depth material about the Higgs mechanism and boson see Lisa Randall's *Knocking on Heaven's Door* (published by The Bodley Head), which includes still more about particle physics, the Higgs mechanism and boson, super-symmetry, extra dimensions, LHC experiments, the nature of science, and the many elements of the scientific process.

www.vintage-books.co.uk
www.bodleyhead.co.uk